It’s good when the pieces stay together.

Penelope, Paige and Paxton are shopping for their school carnival at Puzzleland Emporium.

Published by the Honey Bee Company LLC
2906 Central Street • Suite 258
Evanston, IL 60201
847.864.8032

ISBN 978-1-947933-04-00

My name is:

____

I'm ready to get to work!

They're making sure that Penelope's mom gets everything they need.

What happened first, second, third and last? Number the order of events.

Penelope, Paige and Paxton love Puzzleland Emporium.

How many packages do you see?

I see ________ packages.

It has things to play with.

Circle the smallest one.

It has things to eat.

Circle the largest one.

It has places to sleep.

Color the picture.

At Puzzleland Emporium,
there's always something new to do, to see and to learn.

Start →

2	20	22	24	26
4	18	16	14	28
6	8	10	12	30

End

Help the three friends get to the robot sweeper.
Draw a path counting by 2s.

Today, the three friends learn about packaging from the experts: packaging engineers.

Packaging Engineering

is the way we make sure packaging works the way it's supposed to work.

Packaging works the way it's supposed to work when it covers and protects the things we use.

It's packaging that makes sure that things don't get messed up or broken before we can use them.

How many eggs does the package hold?

The package holds ________ eggs.

Just like engineers design buildings and towers, there are engineers who design packaging.

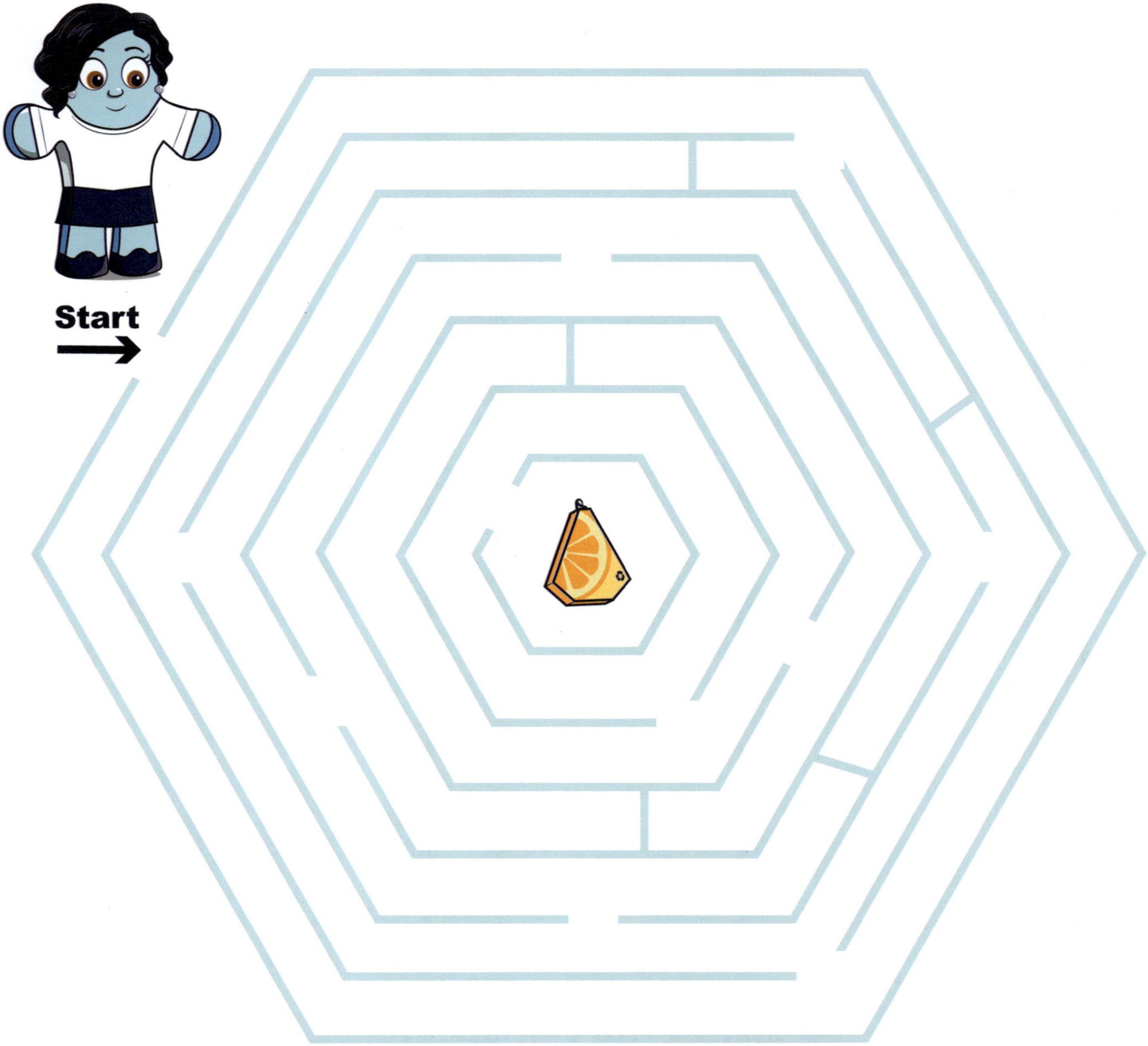

Help the packaging engineer get to her package.

Packaging engineers love to design all kinds of packages.

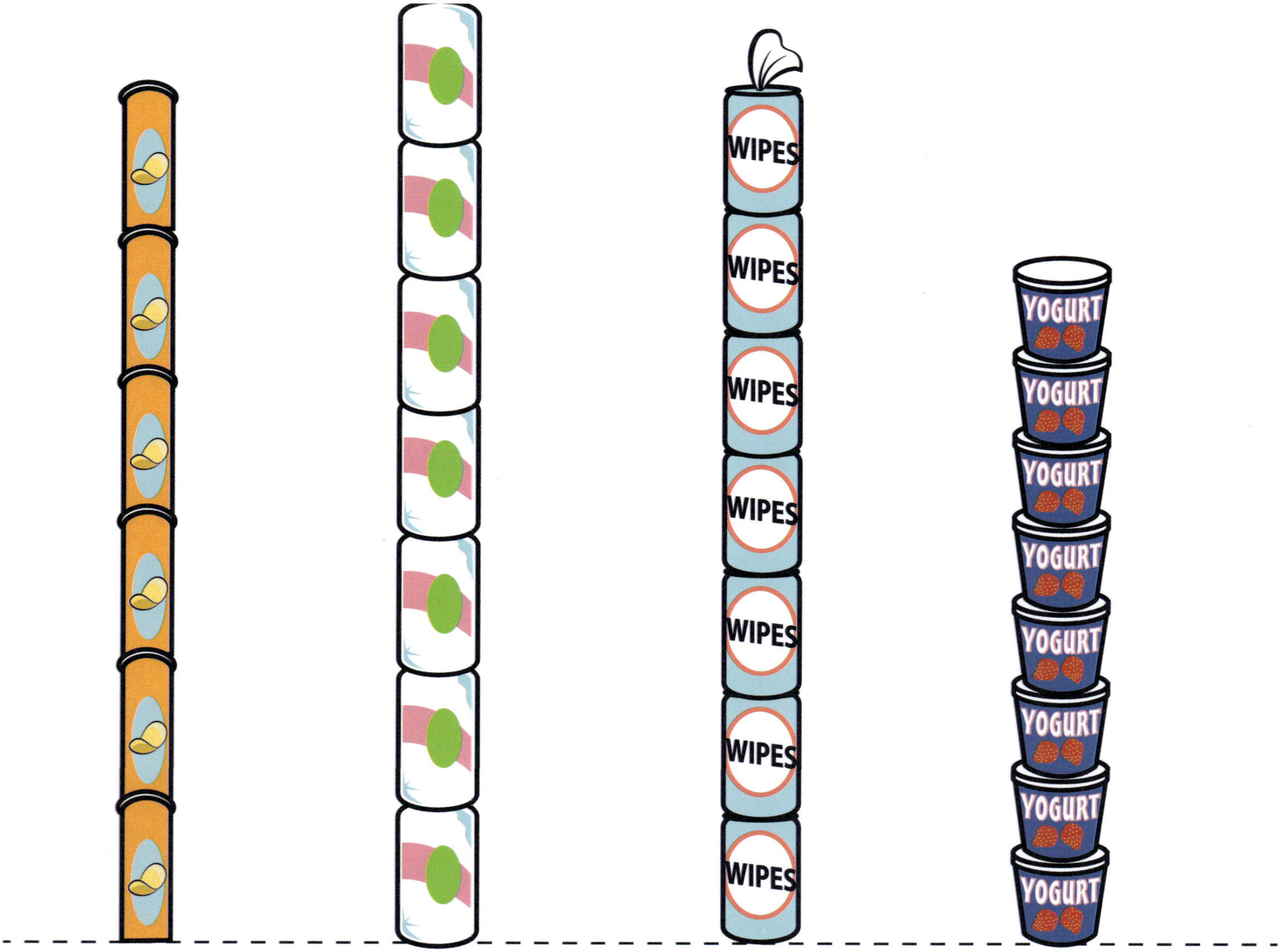

Measure the shortest one.

The shortest one is ________ inches high.

Packaging engineers love to make all kinds of packages.

Draw lines to the ones that are the same,

Packaging engineers love to decide what goes in what type of package

Different kinds of packaging is good for different kinds of things.

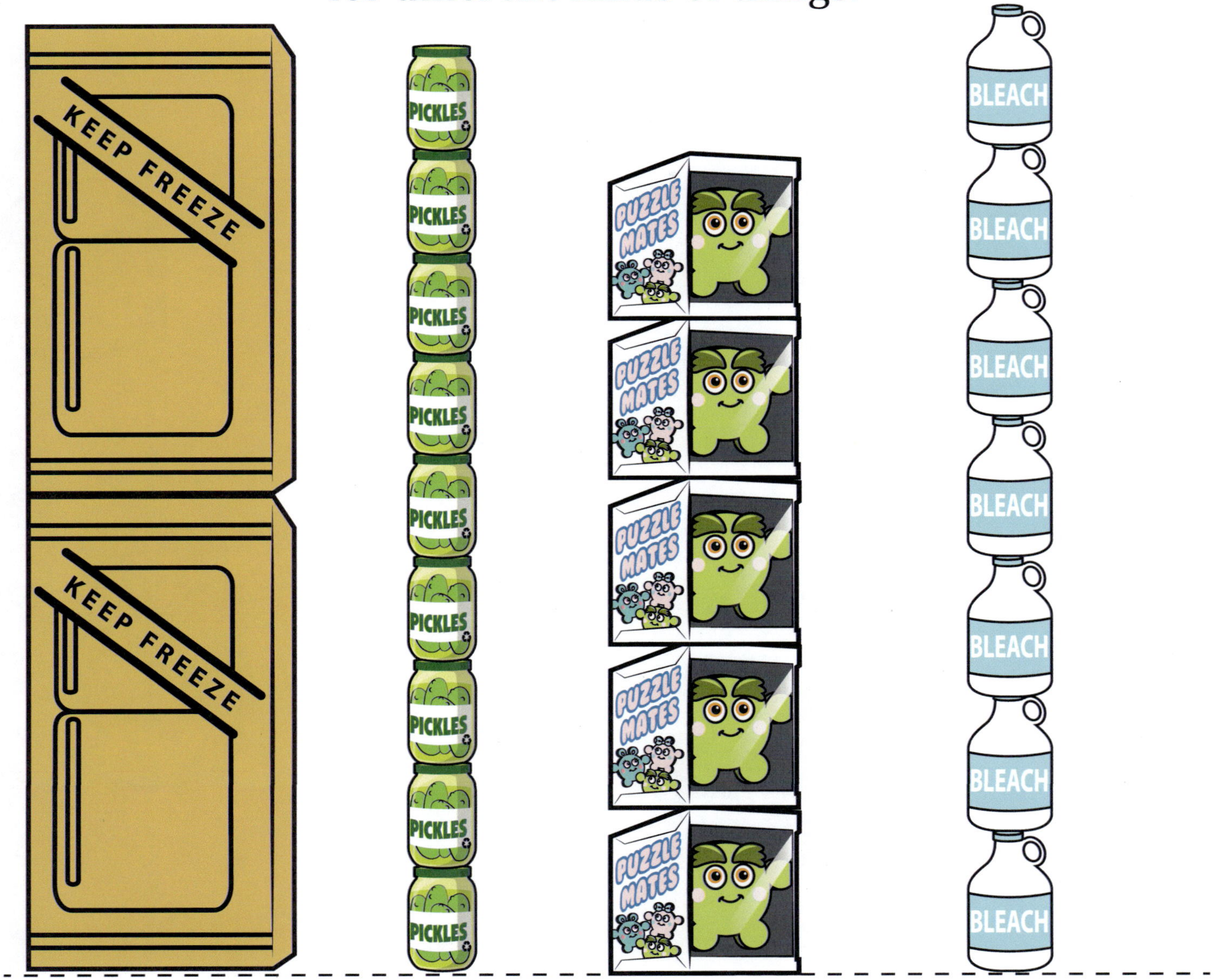

Measure the tallest one.

The tallest one is ________ inches high.

Food is packaged to stay fresh.

Circle the one that doesn't belong.

Electronics are packaged to prevent damage to parts.

Liquids are packaged to prevent leaks and pour easily.

Circle the picture that is different.

Paper is packaged to prevent it from getting wet or dirty.

Draw lines to the ones that are the same.

The best packaging protects the goods, doesn't cost too much and doesn't leave a lot of waste behind.

The best packaging is also

Recyclable

Recycling turns packaging into something else we can use instead of throwing it away.

Are there more yellow or green flying disks left?

There are more ______________________________ flying disks left.

What are some of your favorite things that come in packages?

Draw them here.

Design new packaging for one of your favorite things.

Could packaging engineering be your perfect fit?

Our Mission

Change the course of a nation by making STEM (Science, Technology, Engineering and Math) easy to teach and easy to learn at the beginning of a child's educational experience.

Our Vision

World Domination through STEM

Why We Do What We Do

For Kids: 80% of adults in STEM fields had at least one parent in the field. This means a huge number of our kids are missing out on opportunities that can be life changing.

For Teachers: Classroom teachers face unprecedented challenges to meet expectations and standards. Teachers need time-saving resources that enable them to make small shifts in their classroom lessons to drive STEM interest and create a STEM mindset.

For Our Nation: We're losing the battle. Thirty years ago, 40% of the world's scientists and engineers came from the U.S. Today, it's 15%. STEM fields are where the well paying jobs are—and will be—in the global economy.

Like Us On Facebook

Pin Us On Pinterest

Follow Us On Twitter

Made in the USA
Monee, IL
05 November 2026

67380136R00019